We Are STEAMbitious

By Traci Brown, M.Ed. & Breanna B.
Illustrated by Camille Tinio

We Are STEAMbitious
Copyright, Traci Brown, 2019
Published by PLI Books
FLORIDA

Written by Traci Brown, M.Ed. and Breanna B.
Illustrated by Camille Tinio
Creative Design and Concept by Traci Brown, M.Ed.

All text rights reserved. No portion of this book may be used or reproduced in any manner whatsoever without written permission except in the case of brief quotations embodied in critical articles and reviews.

We Are STEAMbitious

Table of Contents

Welcome to We Are STEAMbitious	p. 1
SCIENCE	**p. 4**
Food Scientist	p. 5
Astrophysicist	p. 7
Forensic Scientist	p. 9
Orthodontist	p. 11
Environmental Scientist	p. 13
Science Activity: Making Slime	p. 17
TECHNOLOGY	**p. 20**
Mobile App Developer	p. 21
Digital Illustrator	p. 23
Coder	p. 25
Video Game Designer	p. 27
Cryptanalyst	p. 29
Technology Activity: Build a YUM-puter	p. 33
ENGINEERING	**p. 36**
Roller Coaster Engineer	p. 37
Inventor	p. 39
Biomedical Engineer	p. 41
Civil Engineer	p. 43
Aerospace Engineer	p. 45
Engineering Activity: Super Sour Pucker Power	p. 49
ARTS	**p. 52**
Writer	p. 53
Musician	p. 55
Photographer	p. 57
Choreographer	p. 59

Fashion Designer	p. 61
Arts Activity: Poetry & Painting	p. 65
MATHEMATICS	**p. 68**
Economist	p. 69
Mathematician	p. 71
Cartographer	p. 73
Actuary	p. 75
Accountant	p. 77
Math Activity: Become a Mathe-magician	p. 81
Draw a STEAM Career of Your Own	**p. 84**
Glossary	**p. 85**
Answer Key	**p. 87**

Welcome to We Are STEAMbitious!

We believe in helping to make the idea of achieving success and accomplishing goals a natural part of every girl's thinking. *STEAMbitious* is about giving girls an achiever's mindset that acknowledges, respects—and places at the center of achievement—hard work, determination, and perseverance. Practice makes progress. And progress builds confidence and feeds motivation. And confidence combined with motivation turn dreams into reality.

We hope that when little girls read the career description and engage in the activities inside of this book, they do so with the confidence of knowing that they can do whatever they want to do in life—whether it's something found in this book or not—and are inspired to do further research on STEAM and other careers that may be of interest to them.

What's Inside:

- Descriptions of 25 different STEAM careers
- *Did You Know?* Fun facts
- Puzzles, games, and practice problems
- Coloring page for each career
- Hands-On Activity at the end of each of the 5 STEAM sections
- Research opportunity for each Hands-On-Activity
- Draw a STEAM Career of Your Own
- Glossary
- Answer Key

How to Use This Book:

You can read this book independently or with your parent, teacher, or friend. Color the pages however you want. There are no rules for your creativity. Have fun!

"Finding the Connections" Challenge:

We challenge you to find the "connections between the sections." What this means is that we want you to see if you can determine how the various STEAM careers relate to one another or where they overlap. For example, the Digital Illustrator career is listed under the Technology section of this book, but it could also be listed under the Arts category because illustrators are also artists. They could even be listed under Math because there is math involved in the digital creation process. Engineering comes into play because of the construction of the device on which they are working. Someone had to engineer the digital tablet. In addition to these other categories, Science is the foundation for technology and engineering, so Digital Illustrator could have been placed under this section as well.

Share the fun of the challenge by getting a grown-up in your life to work with you. Challenge them to see how they can explain the connections between the different sections for each career. Also, while doing the hands-on activities at the end of each section, please get a responsible adult to help you.

SCIENCE

FOOD SCIENTIST
(Food SY-n-tist)
"I make science tasty!"

Mmmm . . . Do you smell that? It smells yummy. Let's taste it. Delicious! You may be surprised to know that some of our favorite foods and flavors—macaroni and cheese, sour cream n' onion potato chips, chocolate pudding, frozen pizza—are created in a laboratory by a special kind of scientist.

Food scientists are scientists who use **chemistry** to create exciting flavors in our foods. Chemistry is the study of **matter**, which is anything that has mass and takes up space. Matter itself is made of atoms, which make up everything in the universe, including you, the stars, and even this book.

Food scientists use their knowledge of chemistry to mix different chemicals to make delicious food flavors for us all to enjoy. In fact, some of our favorite treats, such as cotton candy and bubble gum, are invented by food scientists as opposed to Mother Nature. Mix a little bit of this with a little bit of that, and—*Viola!*—pumpkin spice marshmallows. They're like food magicians. They can really trick us. They can even make us believe that there is a big bowl of buttery popcorn somewhere in the room with us when it's really just a beaker full of chemicals we're smelling. No *pop!* No *corn!*

These cool chemists don't just work with food aromas. They also create fragrances for perfumes and scented candles. That apple pie candle may smell just like grandma's homemade special, but there aren't any apples, cinnamon, or sugar in that jar full of wax. Same with that piece of strawberry-flavored gum you chewed earlier. It didn't have one single bit of strawberry in it. It's all just flavor and fragrance science.

Did You Know... If you've ever wondered why you can't taste your food when you have a cold, it's because your sense of taste is related to your sense to smell. Your stuffy nose won't let you smell your food. Therefore, you can't taste it either.

Unscramble the following for a treat as light as a feather:

NTCOOT AYDCN

_____   _____

ASTROPHYSICIST
(As-troh-FIZ-uh-sist)
"My imagination is bigger than the universe!"

Look at the stars twinkling in the sky. Some of those splendid lights aren't stars at all but planets. Have you ever seen the moon through a telescope? Do you ever wonder how many just like it exists?

Astrophysicists are the ultimate universal **sleuths**. They search the skies and beyond for answers to cosmic mysteries, using astronomy (the study of stars) and physics (the study of forces and motion). They hunt for hidden clues to find answers to some of our most curious questions about the universe, such as *How old is it? How did it begin? Are we alone in it? What else—or who else—is out there?* They get to use the world's biggest telescopes and point them toward space and see things most of us never will. They can even use them to see back in time!

According to NASA, the goal of an astrophysicist is to "discover how the universe works, explore how it began and evolved, and search for life on planets around other stars" (yourfreecareertest.com). They study moons, planets, stars, meteors, meteorites, comets, galaxies, and even black holes—which are huge dark holes that form when a star explodes. They're like the hungriest vacuum cleaners ever, sucking up everything that comes near—including light! *Yikes!*

Did You Know... NASA is an acronym for National Aeronautics and Space Administration. They build and place telescopes into space so that astrophysicists can study what's "out there."

Cosmic Hunter

Most of what exists in the universe are invisible to our human eyes, so we call it "dark matter." Can you use the word bank below to find the hidden words?

WORD BANK

STAR
PLANET   COMET
BLACK HOLE   MILKY WAY
SATELLITE   UNIVERSE

```
A B J Y S X T X R T
Y L R A T S E V W E
P A M E E R N O T U
A C I C Y P A I F N
V K L O O Y L C B I
V H K M P L P G B V
B O Y E E L A X P E
S L W T F A W R N R
L E A Y G O F X X S
N S Y X P G W I Q E
```

FORENSIC SCIENTIST

(Fuh-REN-sik SY-n-tist)

"My experiments help make society safer!"

Oh, boy! Who took the last cookie that was promised to you? If you were a forensic scientist, you would know just what to do to solve this mystery and crime against your appetite. One of the first things you might do is dust the cookie jar for fingerprints since everyone's fingerprints are unique. That's right. No two people on the planet have the same fingerprint pattern. Therefore, this evidence you'd collect would narrow down the list of suspects to just a single person.

Forensic scientists are scientists that help to solve crimes using—you guessed it—science! Only rather than figuring out who took the last cookie, they solve real-life crimes, serious ones that can help put the bad guys and gals behind bars. Their discoveries are used in court to assist juries and judges in making decisions on whether a person is guilty or innocent. That means the work forensic scientists do is pretty important.

It takes the work of several different professionals to help solve a crime. Detectives observe a crime scene, and crime scene technicians take pictures and collect evidence from the scene, such as fingerprints, loose fibers from carpet or clothing, and even **DNA** samples from sources like hair and saliva. (DNA is the stuff inside your cells that holds the instructions on how to make you YOU.) This evidence is sent to the crime laboratory, where a forensic scientist performs experiments on it and then **examines** the results. They use their strong knowledge of science to help determine what happened, when it happened, and, possibly, who did it. Their work helps to make our society a little bit safer by taking dangerous criminals off the streets.

Can you help solve the cookie jar caper?

Which of these suspects' fingerprint is a match to the one found on the cookie jar?

Print found on the cookie jar   **Suspect #1   Suspect #2   Suspect #3**

Write you answer here: _____

ORTHODONTIST
(Or-thuh-DON-tist)

"I help you love to say 'cheese'!"

Nothing says, *"Hello World!"* like a beautiful smile. A friendly smile from someone can really brighten a person's day. Sometimes those perfect smiles are the work of one very special kind of doctor: an orthodontist. Orthodontists are doctors that help beautify the world by helping people achieve the smiles that make them feel confident and allow them to shine.

Your mouth is so important that there are several special tools just for it, such as a toothbrush, toothpaste, floss, tongue cleaners, mouth rinse, even water picks. But while a dentist focuses on your **oral hygiene**—how clean and healthy your teeth and gums are—an orthodontist follows up with making sure that those clean, healthy teeth and gums of yours are lining up the best way they possibly can.

Braces are dental appliances that an orthodontist may put onto your teeth to make them straighter. They look like railroad tracks going across your teeth. When braces are tightened every few weeks, they help set your smile on the right track. Even adults get braces because you're never too old to have the smile that makes you happy.

Did You Know... An X-ray can show not only your baby teeth but also the ones that are hiding beneath the gumline, waiting to make their appearance. That's right! Your adult teeth are already there, sleeping just below your baby teeth, awaiting their turn to break through the surface. It looks rather weird on an X-ray to see *four* rows of teeth instead of two. They may even remind you of shark's teeth!

Similarities and Differences

How are dentists and orthodontists the same? How are they different? Place the keywords inside the Venn Diagram to show what these professionals are focused on.

cleaning	floss	braces	retainer
X-rays	straightening	check-ups	smiles

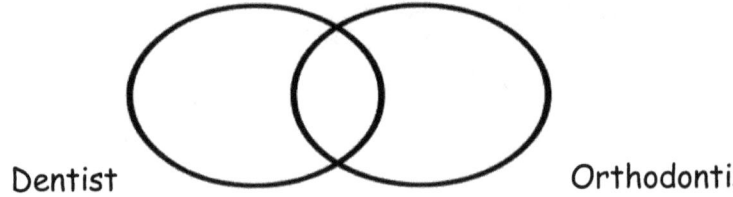

ENVIRONMENTAL SCIENTIST
(En-VY-ruhn-mental SY-uhn-tist)
"I help take care of our precious planet!"

What a beautiful home we have on our planet! The flowers and trees, the cute animals and lovely butterflies all help to make the Earth a wonderful place to live. It's a good idea to remember that all these living things have something very important in common with us humans: they need water in order to stay alive. As important as food is, even *it* can't grow without water. That means we need to take good care of the water on our planet.

Environmental scientists, such as water conservationists, are very special types of scientists that make it their mission to care for the planet. Part of what they do is making sure that our water is clean and free from **pollution**.

Mother Nature does a wonderful job of cleaning our planet's water. She uses the water cycle to do this. However, sometimes we don't always take care of our planet the way we should, and when this happens, our water supply can suffer, and not even the water cycle can fix it. Oil spills, toxic waste, and litter can contribute to turning our wonderful, clean water into some pretty icky stuff! One way we can help environmental scientists protect our planet is to dispose of our trash responsibly, and to turn off the faucet when we're not using water.

Did You Know... All the water on Earth right now has been here since Earth's very first birthday. *Day One.* That's super amazing since Planet Earth is 4.5 billion years old. If we threw Earth a birthday party, we'd need 4,500,000,000 candles!

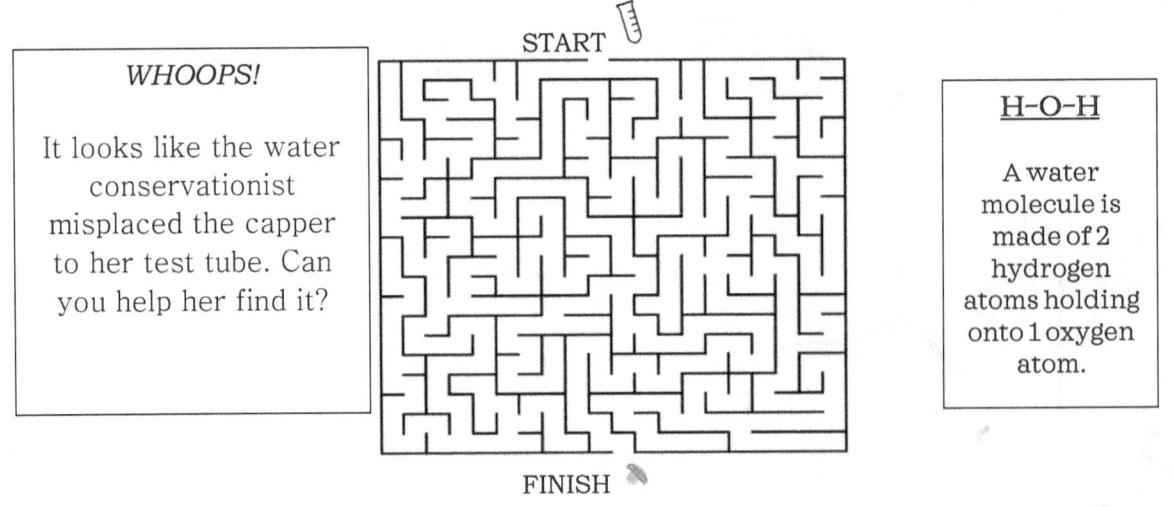

Science Activity

Making Sparkly Slime

Materials:

- ½ cup of liquid starch
- 1 bottle of glue
- measuring cup
- mixing bowl
- large spoon or rubber cake spatula
- gloves
- food coloring or markers (your choice of color)
- glitter
- jewels or rhinestones
- an adult to supervise

Procedure:

1. Use measuring cup to measure 1 cup of glue.
2. Pour a cup of glue into the bowl.
3. Add a few drops of food coloring to the bowl.
4. Stir to create a mixture. (If you need to add more food coloring you may do so until the color is just how you want it.)
5. Little by little, slowly pour the liquid starch into the bowl's mixture. Make sure to stir the entire time.
6. Remove the slime mixture from the bowl and use your hands to knead it. You will see and feel it thicken.
7. Place the slime back into the mixing bowl.
8. Sprinkle some glitter into the slime and knead it some more.
9. Sprinkle some jewels or rhinestones into the slime. Knead it more.
10. Enjoy your sparkly slime!

Become the Researcher

Slime is so fun to make and play with. But how does it work? What's the science behind the slime? Become the researcher as you seek to find out. Record your findings in the space below.

TECHNOLOGY

MOBILE APP DEVELOPER
(MOH-bill AP Dih-VEL-uh-per)

"My new app is fun at your fingertip!"

Phones, tablets, iPads, and desktop computers are necessities in our society. We use these gadgets to gather information, stay up to date on current events, communicate with friends and family, watch movies and videos, learn about the world, and even study a new language. Our favorite part of our electronic devices is the apps, which is short for *applications*.

Imagine getting a new bike for Christmas and then discovering that it doesn't have pedals. Well, that's just as strange as a phone or computer that has no apps. Apps give us access to the cool stuff. They help us go places. Without them, a phone is just something to call someone with, and tablets or iPads wouldn't do anything at all.

There are so many different types of apps. There are game apps that include word searches, makeover challenges, and dance battles. There are apps for social media platforms such as Twitter and Instagram. There are apps for taking and editing photos and videos, listening to music, watching movies, shopping, ordering food for delivery, creating graphic designs, reading or listening to books, checking in at the doctor's office, getting directions to someplace special, checking the weather, monitoring your heart rate during exercise—everything! Whatever needs to be done, apps make life easier, more convenient, and more fun.

Did You Know... The first cell phone was sold in 1983 and cost $4000.

If you were a mobile app developer, what kind of app would you develop? (It might help to think of something you would like to have or believe you need.)

Create an App of Your Own	**Draw your app here:**
Name of your app: _____ Purpose (What does it do?): _____ Target audience (Who is it for?): _____ How it works: _____ _____	

DIGITAL ILLUSTRATOR
(DIJ-i-tul IL-uh-stray-ter)
"My art comes to life on the screen!"

Art is all around us, giving our eyes something interesting to look at and stirring our hearts and imaginations. Think of some of your favorite books or magazines. Do you have a cool movie or character poster on your wall? One of more digital illustrators put their talent to work to give you those amazing visual experiences!

Digital illustrators are artists that create illustrations such as drawings, paintings, and sketches by hand using digital tools rather than using a traditional pen, pencil, or paper. Instead, they use a mouse, pointer, digital stylus, and digital paintbrushes with a tablet or iPad to create their images. They can also use these digital tools to create special effects, such as giving their images the appearance of having a glow or soft shadows, or even giving it a look of texture like stone, skin, velvet, and fur.

Digital illustrators create artwork for print and digital display. They help to create images for books, newspapers, magazines, video games, websites, posters—tons of things! And they do it all from scratch! It takes practice for them to get used to doing it this way, but once they do, their creations are mind-blowing.

Did You Know... Photographs and digital illustrations are made up of pixels. **Pixels** (short for "picture elements") are tiny, colored, individual dots or squares that, when put together, make up a picture. Thousands—even millions—of pixels are used to make a single image. So, when you think you're looking at one object, you are actually looking at *several* tiny objects that have come together to make it look like one.

Guess the Number of Pixels

Can you guess the number of pixels that make up this scientist?	How many pixels do you guess? _____

CODER

(KOH-der)

"I tell computers what to do!"

Computers are everywhere. They are inside of our phones, watches, cars, airplanes, game consoles, hoverboards, washing machines, and much, much more! Pretty much anything electronic these days has a computer inside of it. They do so much for us, and they're great at taking orders. But before they can take our orders, someone must program them to follow instructions. Who gives computers instructions on how to follow *our* instructions? Coders do, that's who!

Coders, also known as computer programmers, take the program designs from a **software developer** or **software engineer** and turn them into a language that a computer can understand and follow. This is called writing computer software or, as we like to say, "writing code." A code is a set of instructions that a computer can understand. A tourist may need a translator to translate from English to French in order to understand what is being said to them while vacationing in Paris. Likewise, computers have their own languages, such as Python, C++, Pearl, and Ruby, that also require translation. A coder takes the software developer's instructions and translates them into one of these computer languages. These instructions must be specific and step by step. Otherwise, your striking keys on your device won't mean a thing. Without a code being written for it, a computer doesn't know how to play a video, send a message, or what to do when you tap an app.

After a coder programs a computer, they test it to make sure there are no errors (or **bugs**). This ensures the apps and programs can function properly so that our digital toys work as they should. Coding is simple and easy to learn. Kids learn how to do it every day!

Did You Know...The first computer programmer was a woman. Her name was Ada Lovelace.

Write in **HTML** (Hyperlink Text Mark-up Language)

Write the following sentence in HTML language on the line below:

I love **STEAM**!

HTML Cheat Sheet
- `<b>` - start bold
- `</b>` - stop bold
- `<i>` - start italics
- `</i>` - stop italics

_____

VIDEO GAME DESIGNER

(VID-ee-oh GAYM Dih-ZY-ner)

"I design so you can win!"

If you've ever visited an arcade, you know that there are tons of fun and exciting video games to play. From color matching puzzles and racecar driving to virtual paintball and castle takeovers, video games can be an exciting way to pass some time and to engage with someone else in a little friendly competition.

Video game designers are the brains behind the buttons. They use their knowledge of computers, computer programming languages (ex. **JAVA**), and special computer **software** to create games that bring tons of screen fun to children and adults everywhere.

It takes teamwork to make a game great. Therefore, video game designers work with artists and computer programmers on things like concept, setting, characters, and story. This means that, together, they decide what the game will look like, who the characters will be and how they will look, what the challenges and storylines will be, what graphics will be used, the rules and goals of the game, and much more in order to deliver to your living room, game room, or hand-held device the perfect gaming experience.

There are so many types of video games out there. Sometimes, the best are the educational ones. But no matter the theme or concept, with an exciting career like this one, the game's never over.

Did You Know... Pac-Man was one of the earliest video games created nearly 40 years ago. It was created in Japan and released on May 22, 1980.

Grab a friend and play "Pac-Man Tic-Tac-Toe"! (One of you can be Pac-Man, and the other can be the ghost.)

Mix 2 Classics!

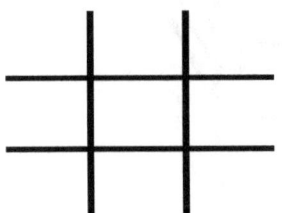

Choose your character:

CRYPTANALYST
(Kript-AN-uh-list)
"I can crack the code!"

Have you ever written with invisible ink or used a secret code to communicate with a friend? Do you have a passcode on your digital diary that no one else knows in order to keep snoopers out? Well, if you happen to forget your own passcode, a cryptanalyst can help you regain your access to your precious, private thoughts.

Cryptanalysts are computer people that can crack almost any secret code. They are good at decoding encrypted information. Encrypting is when you make text private. Cryptanalysts have the ability to **decrypt** the text and read what's being said.

These days everyone has passwords, locks, and secret codes on their personal, digital items to keep their information safe and private. Privacy is a wonderful thing. However, sometimes it may be important to crack certain codes in order to keep our country and world safe. Therefore, cryptanalysts often work in cybersecurity jobs ("**cyber**" relates to computer stuff, like the internet) for the government or for private companies that need to protect their computers, networks, and systems from "digital attacks" carried out by computer villains like hackers, thieves, scammers, or any other person who means to do harm. Whatever the secret plot may be, luckily, when cryptanalysts are at work, a secret is not so secret after all.

Did You Know... Julius Cesar used a **cipher**, or secret code, for his messages. He shifted each letter 3 places to the left. So, D became A, and B became E, and so on, and so forth.

Your Turn: Can You Decrypt This Message?

Use the key to the right to decode the following:

#	A
@	D
$	G
☺	R
◇	M
♀	E
∞	I
☆	B

@ ☺ ♀ # ◇   ☆ ∞ $

Technology Activity

Build a YUM-puter!

For this activity, you will build a computer made of cake and candy!

Materials:

- cake batter w/all ingredients from box
- milk chocolate, white chocolate, and/or peanut butter chips
- fruit roll up
- frosting
- white candies
- edible marker
- Kit-Kat or Twix chocolate bar
- cardstock
- marshmallow fluff
- Twinkie or Sno Balls
- vanilla wafers
- Twizzlers ropes
- Hershey bar (individual squares)
- square cake pan or loaf pan
- parchment paper
- scissors
- 3 wooden skewers
- an adult to supervise

Procedure:

1. Mix cake according to box recipe.
2. Add chocolate chips or peanut butter chips to the cake batter.
3. Bake cake according to box recipe.
4. Spread frosting on the cake and let cool.
5. Using an edible pen, write a different letter on each piece of white candy.
6. Use a toothpick to place a tiny bit of marshmallow fluff on each row of the Hershey bar. Make sure the number of marshmallow fluff dots corresponds with the number of letters that will go on each row. (Ex. the first row will need 10 dots of marshmallow fluff.) Space the dots out to ensure there is enough room for each letter on the row.
7. On the Hershey bar, place the candy letters on the three rows to match a keyboard (QWERTYUIOP on the first row, ASDFGHJKL on the second row, and ZXCVBNM on the third row).

8. Use marshmallow fluff in order to stick the first Twizzlers rope to the candy bar keyboard.
9. Unwrap Twinkie or Sno ball and attach the Twizzlers rope to the top. You can use marshmallow fluff to hold the rope inside if it doesn't fit.
10. Break one vanilla wafer in half to make the "ears" of the mouse. Using marshmallow fluff, stick half of the wafer cookie on both sides of the Twinkie or Sno Ball.
11. Unwrap a fruit roll up and adhere to parchment paper. Use marshmallow fluff to stick the entire sheet to the cardstock.
12. Affix cardstock to a piece of cardboard.
13. Use three wooden skewers to prop up the fruit roll up monitor on top of computer CPU.
14. Attach keyboard Twizzlers end to the front side of the cake, and the mouse Twizzlers end to the side of the cake.
15. Grab a friend and name all the parts you just assembled. Then, enjoy!

Become the Researcher

An edible computer is an unusual and yummy treat. But what are the *real* parts of a computer? And what do they do? Become the researcher as you find out. Include sketches as you record your findings in the space below.

35

ENGINEERING

ROLLER COASTER ENGINEER
(ROH-ler KOH-ster En-juh-NEER)
"My rides give you a thrill!"

Amusement parks are packed with people looking for fun. *Scary* fun. *Fast* fun. *Memory-making* fun. Roller coaster engineers are the makers of that fun. They use the science of **physics** to create a time to remember.

There are different types of roller coasters: wooden coasters, steel coasters, and themed coasters too. They can take anywhere between 8 months to 5 years to complete, and they cost about $8,000,000 to build. (That's *millions!*) It takes several different kinds of engineers to design the fun and excitement that riders crave. These engineers use the teamwork it takes to build rollercoasters that will guarantee both fun and safety. Mechanical engineers work on the machine parts. Electrical engineers connect and power things to move. Structural and industrial engineers use gravity, **velocity**, and the flow of water to reduce friction and to make things slide downward from up high. Civil engineers handle things that relate to energy being used and needed. Computer engineers use computer technology to calculate friction, mass, rider spacing, and curve size. There are even design engineers who build artificial rockwork and props that create natural-looking settings such as logs, trees, wildlife, paintings, and carvings, which help to make things look pretty and **camouflage** all the mechanical parts and other things we don't need to see.

Wow! Who knew so much went into a rollercoaster ride? Every twist, every turn, every drop, every loop was thoughtfully and carefully designed to provide a specific thrill and to create a certain effect on the human body. Engineered excitement!!!!

Did You Know... The world's tallest roller coaster is at Six Flags Great Adventure in Jackson, New Jersey. It's called *Kingda Ka*, and at 456 feet with a 418-ft drop, it's sure to give riders the stomach-dropping thrill of their lives.

Complete the following Awesome Analogies

Wheels are to roll as wings are to _____.	Rollercoasters are to up as submarines are to _____.
Pool is to swim as trampoline is to _____.	Popcorn is to movies as cotton candy is to _____.

INVENTOR

(In-VEN-ter)

"My toys are machines of fun!"

The light bulb. The phone. Cars. Clocks. These are inventions that have changed the world. It's hard to imagine life without them. Inventions like these make life easier. Inventors like Ruth Graves Wakefield and Sue Brides, who invented the chocolate chip cookie, make life yummier!

Inventors make cool things. They're not afraid to take stuff apart in order to figure out how it works and then use this knowledge to make even more amazing things. Many use their knowledge of one or all STEAM fields to develop their products.

So many wonderful inventors have given us joy and convenience through their creations. Nancy Johnson has earned a special place in every kid's heart as she invented the ice cream maker. Babies and teachers around the world are glad that Adeline D. T. Whitney invented alphabet blocks. And anyone who's ever been on a sinking boat has Maria Beasley to thank for inventing the life raft. Florence Parpart, who invented the electric refrigerator, is as cool as her invention. Eco-friendly inventor, Maria Telkes, invented the first house completely powered by the sun. If you like board games, give a shout-out to Elizabeth Magie, who invented the first Monopoly game. For those of us who love geography, we have Ellen Fitz to thank for inventing the globe. Jeanne Villapreux Powers deserves to have every aquarium named after her, especially since she invented them. And millions of parents, as they tuck their children into bed at night, can give a special thanks to Mrs. Maria Van Brittan Brown for inventing the first home security system.

Did You Know... Research chemist, Stephanie Kwolek, invented Kevlar, which is a **synthetic** fiber that's five times stronger than *steel* and is used in hundreds of things from bicycle tires and frying pans to musical instruments and snow skis.

Can you match the inventor with the invention?

Nancy Johnson	solar-powered house
Elizabeth Magie	ice cream maker
Maria Telkes	aquarium
Jeanne Villapreux Powers	Monopoly game

BIOMEDICAL ENGINEER
(By-oh-en-juh-NEER)
"I help make life a little better and easier for my fellow human beings."

You'll take lots of pictures in your lifetime. Your very first picture, called a **sonogram**, was taken before you were even born using a special machine called an **ultrasound**. Using sound waves, it peeked through your mommy's tummy to find you sleeping peacefully inside. This, X-ray machines, and several other inventions were created through biomedical engineering.

Helping one another is one of the best uses of our time. Sometimes people have physical challenges. Maybe they have lost a limb (an arm or a leg). Perhaps they have a heart that doesn't work as well as it should. Biomedical engineering is a field of technology that assists doctors and nurses with taking care of their patients by finding new and creative ways to address their challenges. Bioengineers create machines, devices, and other technologies such as artificial limbs. Artificial limbs are called **prostheses**, and are made of plastic and metal, and can feel just like a real limb for the person who needs it. Pacemakers are tiny machines that go inside of a person's heart to help it beat better. If you know someone who has ever broken a bone, metal pins were used to hold it in place while it healed. Even your grandpa's dentures were created by a biomedical engineer.

Sometimes accidents can happen. Other times people are just born with challenges. While these things can make us feel sad, it's really good to know that there is help available for these strong and brave human beings through biomedical engineering.

Did You Know... Eye laser surgery, a biomedical engineering invention, eliminates the need for eyeglasses.

Can you identify the body organs in the crossword? (Hint: one is not an actual "organ.")

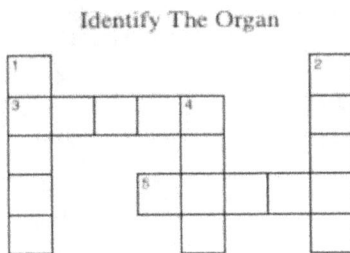
Identify The Organ

<u>Across</u>
3. Helps you breathe
5. Cleans your blood

<u>Down</u>
1. Carries oxygen, nutrients, and waste
2. Pumps your blood
4. Protects your body

CIVIL ENGINEER
(SIV-uhl En-juh-NEER)
"I help build cities!"

Skyscrapers are a sight to see. How in the world can such a tall building stand without tipping over? For the answer, you'd have to ask a civil engineer.

Civil engineers are engineers who figure out the best ways to build **infrastructures**. Infrastructures are things like buildings, railroads, and bridges. Civil engineers design these things and more. They design domes and towers that attract tourists, dams to keep water at bay, and tunnels and bridges to go over and *under* water— even between mountains! Civil engineers know all about concrete, asphalt, and steel, and they know what a building should be built on top of and what it should not. (This is called a building's **foundation**.) Civil engineers make sure things are built safe and strong.

Civil engineers also design our roads. They help make traffic run smoother by figuring out the best way to safely get every vehicle on the road to their destinations in the least amount of time possible. In addition, they design airports, our water supply systems, and even our sewage systems that get rid of our waste.

Besides designing, civil engineers also oversee the construction of their projects, making sure things are built correctly according to their designs. They may work at a construction site or in an office. They may even get to travel to different countries to do their work.

Did You Know… Civil engineers help design water slides by telling the slide builders just how much water is required to help us slide down without getting stuck.

Which Does NOT Belong? Which is NOT designed by a civil engineer?

AEROSPACE ENGINEER
(AIR-oh-spayce En-juh-NEER)
"My dreams take flight!"

Look up in the sky and you don't see only birds and clouds. You just might see an airplane! Airplanes really make travelling fun. Have you ever been on an airplane?

Aerospace engineers design, develop, and assemble aircraft and spacecraft. Aircraft are vehicles like airplanes and helicopters. They fly through the air and are used to carry passengers, mail, and other **cargo**. Spacecraft include vehicles or machines such as satellites, missiles, rockets, and space shuttles. They are used for a variety of purposes, such as observing the Earth, studying weather, communications, navigation, exploring other planets, and many other things. Both types of craft must be designed specifically for their purpose and destination. For example, since there is no air in space, airplanes can't fly there. They need air to flow over and under their wings in order to hold them up. On the other hand, rockets must carry their own air with them in order for their engines to work. Aerospace engineers know how to **differentiate** designs for the two.

An aerospace engineer's very first design of a craft is called a **prototype**. (Proto means "first" in Latin.) Once they build the prototype, they test it out to make sure that it works properly and assist with the navigation design. They even help to design airports. Aerospace engineers are involved in several steps of many processes, making sure things fly, soar, zoom, and thrust as smoothly as possible.

Did You Know... When you see the prefix *aero*, it always has something to do with air, oxygen, or gas. The best bicycles and helmets are designed to be more **aerodynamic**, which means that bicyclists are better able to cut through the force of wind resistance, making them go faster.

*** WORD CHALLENGE ***

How many words can you make out of the word **AEROSPACE**? List them below:

> For example, **ROSE**

Engineering Activity

Super Sour Pucker Power

For this activity, you will make a battery out of a lemon.

Materials:

- 1 lemon
- (1) 2-inch long copper nail, screw, or galvanized nail
- (1) 2-inch zinc nail, screw, or galvanized nail
- holiday light with a 2-inch lead (enough wire to connect the two nails)
- an adult to supervise

Procedure:

1. Gently press down on the lemon and roll it around on a table to loosen up the juice inside and soften it up. Careful not to break the skin.
2. Insert both nails into the lemon, two inches apart, making sure they are not touching each other. Be careful not to let the nails go all the way through to the other side of the lemon.
3. Remove insulation from around the holiday light's lead to expose the bare wire underneath.
4. Wrap the exposed wire around both nails and watch your bulb light up!

Become the Researcher

Electricity is amazing! But how does it work? Research the topic. Be sure to learn all about atoms and their subatomic particles (most importantly, electrons). Record your research in the space below.

ARTS

WRITER
(RY-ter)
"I make characters come to life!"

The world of imagination can be the spice of life. Be it through stories, songs, theatrical plays, or poetry, experiencing a world of make-believe and fantasy can add so much joy and entertainment to our lives. Everything that exists in our world began with an idea. It began with someone's imagination.

Fiction writers are highly imaginative people that use their minds to create new worlds and make the unbelievable *believable*. They conjure up characters and creatures and ideas with their imaginations, then write them down to share with the world. They give birth to the characters that become real and important to us. Characters that did not exist before the writer put pencil to paper.

Nonfiction writers use their gift of writing to inform and educate us about different things. They inspire us and take us on journeys based on real-life matters and topics. They teach us about art or science. They teach us about important people in history or special places around the world. Some writers even write about what's happening in the world right now. These writers are called **journalists**.

All writers, fiction and non-fiction, put words together in a way that makes us see, feel, hear, taste, and touch a place in time, real or imagined, all through the power of the printed pages of their work. Some of the best writers make us feel as if we're part of their stories or that they are speaking directly to us. Still, others make us forget about ourselves altogether and take us deep into a world of fantasy and exploration. They make us spectators to someone else's life. They make us care about each other. That's the magic of writing.

Did You Know... The current record holder for the world's tiniest printed book is a children's book called *Teeny Ted from Turnip Town*, measuring 70 x 100 **micrometers**.

 Put on Your Writer's Crown

Finish this opening line:

*L*ong ago, in a faraway land, _____
_____.

MUSICIAN

(Myoo-ZISH-uhn)

"Creating cool beats to move your feet!"

Music is powerful. It has the power to control our feelings and sometimes even our actions. The power of music is something we often take for granted. Music can inspire so many feelings and emotions. It can make us feel happy, excited, understood, courageous, or even sad. Music is often a big part of our celebrations, such as birthdays, weddings, and graduation ceremonies. We hear music at holiday parties, parades, at church, driving in our cars, or just listening to music on our phones or other electronic devices while we chill out.

Musicians are creative and inspiring individuals who have the desire and the gift for making the music that becomes the soundtrack of our lives. There are as many kinds of musicians as there are different kinds of music. You have composers, recording artists and songwriters, to name a few. Like musicians, music also comes in many varieties or **genres**. There is classical music, folk, pop, country, jazz, spiritual, rock, rap, alternative, the blues, bluegrass, and much more. There is also music specific to different cultures and parts of the world.

What does a tuba and French horn have in common? They're in the same family. Yep! Musical instruments have families, too. These families consist of Percussion, String, Keyboard, Brass, and Woodwind. We can also make music with our bodies by clapping our hands or stomping our feet in order to create a special beat. Some musicians use nothing but their mouths to make a melody. Whatever their instrument, all musicians make the world a sweeter-sounding place to live in.

Did You Know... Xylophones, which are part of the percussion family, are often used to imitate the spooky sound of bones clanking in Halloween-themed music.

Can you match the musical instrument with its family?

Piano	String
Triangle	Woodwind
Flute	Keyboard
Trumpet	Brass
Guitar	Percussion

PHOTOGRAPHER
(Fuh-TOG-ruh-fer)
"I can share a moment in a snap!"

From movies and magazines to websites and wall art, photographers give us the world through their magical lenses. A beautiful picture can take our imaginations on the sweetest journey or even inspire us to go on a real-life journey. Those who use cameras as their tools for capturing or creating art add joy and beauty to the world.

Since cameras are everywhere—our phones, in stores, at traffic lights—it can be easy to forget just how special photography is. What is so special about a photographer is how they use their cameras in unique ways to offer us a part of the world that we may never see otherwise. They can travel all around the world, discovering breathtaking sights, eating colorfully delicious foods from faraway lands and distinct cultures, then bring all the world's wonders back home to us via their photographs. Through their photographs, they bring the world to *us*.

Photographers can work on projects such as movies or documentaries. They can shoot for magazines, newspapers, and for companies that need to advertise. A photographer can even choose to specialize in a specific field of photography. There are food photographers, fashion photographers, landscape photographers, travel photographers, special event photographers, sports photographers—all kinds! Through their photographic recordings, each can freeze time, helping us to experience and preserve precious memories and sights.

Did You Know... Our eyes observe an image, then record it on our **retinas** upside down. It's our brain that auto-corrects it for us. A camera works in a similar way.

What We See...	Goes On Our Retina...	Our Brain Corrects It!

Which of these words look the same upside down as it does right-side up?

MOON   SWIMS   XOXO   MEME

CHOREOGRAPHER
(Kohr-ee-OG-ruh-fer)
"C'mon! Move with me!"

Is there a joyful rhythm in your heart that travels through your body and makes you want to dance? Do you wish to use your rhythmic creativity and talent to share the joy of dancing with others? If so, then a choreographer lives inside of you.

Choreography is the art or practice of dancing and/or designing sequences of movement, which is a fancy way of saying making dance routines. Choreographers are those in charge of teaching groups of people dance moves. Think of your favorite school or Broadway musical. Did you like those awesome, theatrical moves between the acting and singing? You can bet that it was a choreographer who designed those cheer-worthy routines and helped to teach the cast of characters how to do them. Ever been to a live music show or concert with lots of high-energy dancing? Your favorite performer may have been the one on stage, but a choreographer behind the scenes deserves a round of applause for coming up with those cool moves that made you and the rest of the audience jump up and down.

Choreographers work in many branches of entertainment. They choreograph for music videos, movies, musicals, commercials, plays, Broadway shows, Vegas shows, TV talent shows. You name it! If there's dancing involved, a choreographer helps the dancers get those moves just right.

Did You Know… Choreographers don't just design for the stage or dance floor—some design for the ice by choreographing figure skating routines.

Dancing Shoes

Match the dance shoe with the style of dance

A – ballet   B – tap   C – Hip-Hop   D – Latin

FASHION DESIGNER

(FASH-uhn   Dih-ZY-ner)

"I help give you style!"

Isn't fashion fun? It can be super cool to mix and match clothing and accessories to create a certain look for the day—one that matches our mood. When you go shopping and see all those wonderful clothes, shoes, and accessories, it's pretty awesome to know that each and every item is a product of a fashion designer's vision.

Fashion designers are artists who create sketches of clothing and accessory designs. Then they go shopping at various places to select fabric and other materials that they want to use for their designs. Traveling gives them ideas, and they get to really have fun picking colors, materials, and prints for their fashion ideas. Every bead, button, sequin, and zipper placement is decided by the fashion designer.

Many opportunities are available to fashion designers. They can work on movie or television show wardrobes. They can design theater costumes. They can dress singers for music videos, models for fashion shows and photo shoots, and celebrities for their special public appearances. Fortunately, many fashion designers sell their special brand of clothing and accessories in stores as well, so that regular people can buy them too. There is so much to choose from. There's sportswear for getting active, outerwear like jeans, shirts, and blouses for casual days. There is formal wear such as suits and fancy dresses. There's footwear—even eyewear, like cool sunglasses. And don't forget handbags, hats, and scarves!

Some fashion designers get really fancy and design princess-style wedding gowns. Others even design *haute couture*, which is a French word to describe super-expensive clothing that is completely handmade and designed specifically for one single person. This is opposed to *ready-to-wear* clothing, which is what we can buy in stores off the racks and where there are several of the same item available for sale.

Did You Know... Technology is helping to improve the fashion world. For example, we now have clothes that are made to provide us with UV rays protection from the sun.

Help! The fashion designer needs to find 4 consonants and 1 vowel in order to start on her next project. Can you help her by filling in the missing letters of the two words below?

__ E W __ N __   __ A __ H __ N E

Arts Activity

Poetry & Painting

For this activity, you will paint a beautiful landscape using traditional as well as nontraditional painting tools.

Materials:

- ❖ Anything you want, such as paintbrushes, sponges, cloths, corks, new toothbrush, cotton balls, your fingers, etc.
- ❖ Any colors you want. There are tons to choose from.
- ❖ Any canvas you want, such as a large white sheet of paper, a piece of tile, a paper plate, cardstock, a large stone, etc.
- ❖ Your imagination. (This is the most important and unique thing you'll need.)

Procedure by Poem:

No Limits

Close your eyes and think of a place
that's filled with beauty, light, and grace.
Select your canvas and colors too.
Whatever you want, it's up to you.

Grab your tools; a brush is fine.
So is anything else that comes to mind.
Now paint what you see inside your head.
Forget everything someone else has said.
Paint from your own heart whatever you choose.
For this activity, there are no rules.
Once you have your canvas colored, try adding some scenery.
It can be rocks, water, sand, animals, or simple greenery.
For your landscape can be whatever you want it to be.
Place no limits on your creativity.

Become the Researcher

How are colors able to blend the way that they do? Which color pigments were the most expensive to buy a long time ago, and why? What's the difference between oil-based, water-based, and latex-based paint? Research, then record your findings below.

MATH

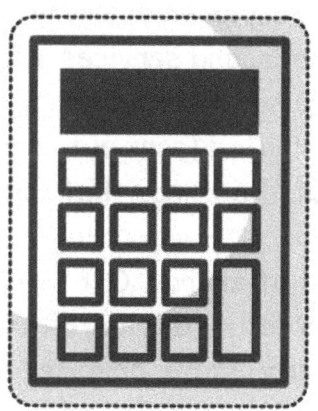

ECONOMIST

(e-KON-uh-mist)

"I can make money your friend!"

It's all about the money with these professionals. If you've got business on the brain, it's good to have an economist by your side. Let's say you were thinking of opening a lemonade stand, an economist could tell you the best place to open it, the best time to open it, and the best price to sell the lemonade for in order to earn yourself enough money to buy those new roller skates you saw. With an economist in your corner, you'll be rolling in no time!

Economists are experts on the **economy**, which has to do with the earning and spending of money in a certain area. In an economy, we are all buyers and sellers. When it comes to **resources**, goods, and services, economists study the patterns of how much of these things are being produced (made), purchased (bought), consumed (used), sold, and distributed (sent out to people) around the world. In other words, they know all about the things that cost or earn us money.

Some economists study the cost of energy. Others study the cost of healthcare, meaning how much it costs to go to the doctor. Other economists study how many people are working in a certain city, state, or other parts of the country or world. They can even study human behavior when it comes to spending money and then use this knowledge to make predictions about what people will buy, when they will buy it, and even *why* they will buy it. This information is very useful to businesspeople, who rely on the word of an economist to make decisions. They're like the fortunetellers of **finance**. They know where the money was, where it went, and where it will probably go next.

Did You Know… Different countries use different things as their type of money or **currency**.

USA	Germany	South Africa	Spain	London	Japan
dollar	euro	rand	peso	pound	yen

Counting Money

If you had (1) ten-dollar bill, (3) one-dollar bills, (2) quarters, (1) dime, (4) nickels, and (7) pennies how much money would you have?

   Total amount = _____

MATHEMATICIAN

(Math-uh-muh-TISH-un)

"Numbers is my middle name!"

Numbers are everywhere. They're on clocks, calendars, road signs, and price tags. They're on thermometers, scales, and money. You use numbers to measure things, describe your age, and to count your trick-or-treat candy at the end of the night.

Mathematics is the science of numbers, quantity, and space. And it doesn't just come in one type. There are many different fields of mathematics, such as algebra, geometry, trigonometry, and calculus. **Mathematicians** are "numbers people" and problem-solvers. They use the magnificence of math to find answers to many of the world's questions. Using equations, they find order in chaos. They make numbers make sense.

There are different types of mathematicians. Theoretical mathematicians create new math rules and theories. A **theory** is a testable explanation based on observations for why things work or how things happen. Applied mathematicians help with solving some fun problems like building space shuttles that fly better or building video game models. Some mathematicians love math so much that they decide to teach math to others. They become teachers at middle and high schools or professors at colleges and universities. They help to create talented future mathematicians who just may decide to use their love and knowledge of math to find a way to build a bridge to Mars! Who knows? With math, there are an *infinite* number of possibilities.!

Did You Know... The first four numbers add up to 10. (Come on, let's add them up!!)

Zero: The Real Hero

Zeros are *super* important. They make numbers larger or smaller. For example, 10 would only be 1 without the zero behind it. Check this out:

------------------------------------- **Big** ------------------------------------- **- Small -**

Hundred	Thousand	Million	Billion	Trillion	One-tenth
100	1000	1,000,000	1,000,000,000	1,000,000,000,000	0.10

Think a trillion is a lot? Well, what about a googol? That's 1 followed by 100 zeros! Feeling brave? Try writing it out across the bottom and up the sides, and all around this page. I'll get you started: 10,000,000,000,_____

CARTOGRAPHER
(Kahr-TOG-ruh-fer)
"I get you places!"

Maps can lead us to wonderful places! They are terrific tools that help guide us toward our dream destinations or even toward hidden treasures. They can help us navigate our way on a nature trail or simply tell us exactly how to get to a classmate's house for her birthday party.

Cartographers make maps for a living. They pair their love of **geography**, art, and design to create surefire steps to get us where we want to go. But map-making isn't just about taking us to where the fun is. It also helps keep us safe and calm in emergency situations. During an emergency, first responders such as firefighters, police officers, and emergency medical vehicle drivers use maps in order to get to those who need help. Meteorologists who report the weather also use digital maps to keep us informed about weather-related events, like hurricanes and thunderstorms. The different colors on the map tell us how strong or weak the storm is in a certain area. Whatever their purpose, having a map certainly makes things easier.

So, next time you're on a road trip with your family, sitting in the backseat, how about volunteering to be in charge of the map? Help to take your family wherever they want to go.

Did You Know... Global Positioning System (GPS) works by receiving radio signals from about 30 satellites orbiting Earth. Three work together to determine your location.

Connect the Dots. What Do You See?

Easter Island is a volcanic island in Polynesia. It's loved by archaeologists for its statues called *moai,* which are carved human figures with oversized heads.

ACTUARY
(AK-choo-air-ee)
"I help plan for rainy days!"

Life is an adventure. At times it can be a bit risky. A **risk** is a chance that things might not end up the way you want them to. Imagine that you plan to spend $30 of your allowance on a particular toy, but the reviews aren't great. They say the toy breaks easily. You can buy insurance on the toy for $4, which means if it breaks, the company will send you a new one for free. You look at your allowance and wonder if the $4 is worth spending. An actuary could help you figure that out.

Actuaries work at taking a close look at past situations to determine how much money should be spent or saved to prepare for a situation that may take place later down the road. They analyze risks and predict the costs needed to pay for certain expected and unexpected events, making sure that money, or some other form of help, will be available when it's needed the most. It's like when your mom tells you to save some of your allowance for a rainy day. Actuaries help people figure out how much that rainy day might cost (should it ever actually rain) and how much you should spend on an umbrella *today*.

There are two main predictions that actuaries make. First, they predict *when* certain random events, like accidents, may occur in the future by looking at past trends. Second, they predict how much money would need to be set aside now in order to pay for that future event that may occur. Many actuaries work for insurance companies, helping them to figure out a fair price to charge their customers for their "rainy day" service. They help to provide peace of mind by making sure that tomorrow's future is well covered *today*.

Did You Know... People spend thousands of dollars every year on insurance in order to protect themselves in case something unexpected happens.

 **Predicting by Patterns**

Actuaries aren't psychics. They just make predictions based on patterns of the past. Can you predict the next series of numbers based on the pattern?

5 * 10 * 9 * 18 * 17 * 34 * 33 * 66 * 65 * ___ * ___ * ___ * ___ * ___ * ___

ACCOUNTANT
(Uh-COWN-tant)
"I will help you manage your money!"

Money can be hard to keep track of sometimes, whether you have a lot of it or a little. Figuring out how much you have earned versus how much you need to spend and how much you've *already* spent can be a little challenging for some people. That's where accountants come in to save the day.

Accountants help others keep track of and manage their money by keeping accurate count of money coming in (earning) and money going out (spending). They keep this information well organized by creating **financial** records that show exactly where the money is and where it isn't. This is helpful to both businesses and individuals.

Many rich people have accountants to help keep up with all that money they have, but regular people can have accountants too. One particular time where many people may seek the help of an accountant is during tax season. Tax season is from January to April, and it's when your parents must report to the government all the money that they have earned the year before. Then, based on that amount, they either get money back (called a tax refund) or have to pay more money to the government. (Here's a tip: you might want to be really nice to your parents around this time of the year if they find out that they have to pay more money. Offer to do extra nice things for them to cheer them up.)

Did You Know... It costs money to spend money. This is called sales tax.

Paying Taxes
(% means "per cent," which means "per every 100." Remember that 100 cents equal 1 dollar.)

If sales tax is 8% in your state, this means that for every dollar (100 cents) you want to spend, you must pay an extra 8 cents in order to do so.

Ex. glitter nail polish @ $1.00 (subtotal) + sales tax @ 8 cents = $1.08 total.

Can you figure out what your total will be for the carnival food items below?

Hot dog - $1.00 + Soda - 50¢ + Chip 50¢   - **subtotal** = _____   **total** = _____

Earnings

chores $5
report card $10
mini bake sale $27
Total: $42

Expenses

new dolls $6
candy $1
lip gloss $2
Total: $9

Balance = Earnings − Expenses
= $42 − $9
= $33

Math Activity

Become a Mathe-magician

For this activity, you will become a *mathe-magician*! You will "mathemagically" be able to correctly guess the number your volunteer subject is holding.

Materials:

- paper or index card
- Sharpies or marker
- calculator (optional)
- five sheets of paper cut into small squares or index cards
- hat
- a stick or baton to be used as the magician's wand
- cape (you can use a hooded jacket, towel, etc.)
- a volunteer

Procedure:

1. Put on your cape and give your volunteer five squares of paper or index cards and a sharpie or marker.
2. Keep one piece of paper for yourself.
3. Turn around, facing in the opposite direction, and tell your volunteer to think of a number—any number! Have them write it down and place it face down. Tell them to make sure to keep it hidden from you.
4. Tell your volunteer to double their number and write down the result on another piece of paper. Again, tell them to place it face down.
5. Tell your volunteer to add 34 to their new number and write down the result. Have them place it face down.
6. Tell your volunteer to divide their new result by 2. Have them write it down and turn it face down.
7. Tell your volunteer to subtract the number that they started with (from Step 3) from the number they have now (from Step 6). Have them write it down and turn it face down.

8. Write the number 17 on *your* piece of paper, fold it, and place it inside of your magician's hat.
9. Turn around to face your volunteer. Show them that there is a paper inside of the hat. (It should be folded well enough so that they cannot see the number you wrote down.)
10. Tell your volunteer to pick up their final card or paper and hold it close to their chest, still hiding the number from you.
11. Wave your magic wand over your hat and say, "Math is magical! Math is magical!"
12. Reach into the hat and pull out your piece of paper.
13. Unfold it and show it to your volunteer, asking, "Is this your number?"
14. They should be surprised when they answer "Yes!" followed by, "How did you do that?!"

Trick Reveal: The final answer is always half of whatever number you tell them to divide by in Step #5.

Become a Researcher

The above activity may *seem* like magic, but it's really just math at work. Ask your teacher, or research on your own, how this "trick" works. Take notes below on your findings.

Draw a STEAM Career of Your Own

GLOSSARY

aerodynamic – dealing with the motion of air and other gases and its effect on things
astronomy – the study of celestial objects, space, and the entire universe
atlas – a book of maps
black hole – super, massive hole in the universe that sucks in everything that comes close
bug – an error or flaw in a computer program
camouflage – to hide, change, or disguise
chemistry – science that deals with matter, including what it's made of and how it behaves
cipher – a message written in code or in a way to conceal (hide) its meaning
computer programming language – instructions given to a computer, including vocabulary and structure, in a language it understands so that it performs specific tasks
cryptic – secret
currency – something that is used to exchange for something else (ex. money)
cyber – refers to computer or computer network, like the internet
decrypt – to break, decode, crack, or decipher a secret code or message
DNA – (Deoxyribonucleic acid) – the molecule that provides instructions on how to make life
economy – the structure and conditions of life in a specific area such as a country, city, or state based on money, jobs, and spending
encryption – a message that is coded in a way to be hidden
examine – to closely inspect
financial – relating to finance
finance – relating to money
foundation – a ground upon which something is built up or overlaid
genre – a category of music
geography – science of the Earth's surface features, including land, life, and culture
haute couture – designers and companies that create one-of-a-kind fashion
infinite – limitless or endless in space, extent, or size; impossible to measure or calculate

infrastructure – the systems and workings of buildings, roads, schools, reservoirs (lakes used as water supply) in a community carried out by the government
JAVA – one type of computer programming language
journalist – a writer, editor, or reporter of the news
matter – anything that has mass and takes up space by having volume
micrometer – one millionth of a meter (also called **micron**)
oral hygiene – keeping teeth and gums clean by brushing and flossing
physics – the science of matter and energy and how they interact
pixel – small elements that come together to form an image
pollution – substances that are harmful and poisonous to the environment
prostheses – artificial devices to replace or improve a missing or malfunctioning part of the body
prototype – the first model of something
resource – something that helps or provides support
retina – the back of the eye that uses light to help the brain form an image
risk – chance or possibility of loss or injury
sleuth - detective
software – one or more computer programs (including apps) that instruct a computer on what to do and how to perform a task, including apps.
software developer/software engineer – builds and creates software and applications
sonogram – an image produced by an ultrasound
synthetic – artificial or man-made
tax – money that is charged to a person for property or a purchased item
ultrasound – machine that uses sound waves to produce an image
velocity – speed over a period of time

ANSWER KEY

Pg. 5. COTTON CANDY

Pg. 7.

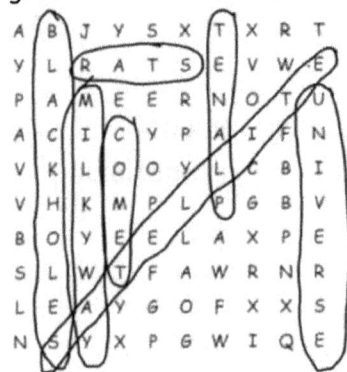

Pg. 9. Suspect #3

Pg. 11.

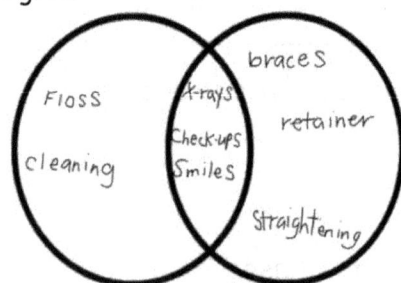

Pg. 13.

Pg. 21. (Your choice)
Pg. 23. 117 pixels
Pg. 25. I <i>love</i> <b>STEAM!<b/>
Pg. 27. (Your game)
Pg. 29. DREAM BIG
Pg. 37. fly, down, jump, carnival
Pg. 39. Nancy Johnson, ice cream maker/Elizabeth Magie, Monopology game/Maria Telkes, solar-powered house/Jeanne Villapreux Powers, aquarium

Pg. 41.

Identify The Organ

Pg. 43.

Pg. 45.
ROSE, SPACE, PACE, ROPE, CAPE, RACE, RACES, CAP, CAPE, CAPES, COP, CARE, CAR, ACE, SEE, SPARE, PEACE, PEA, PAR, ROE. CUP, OAR, SOUP, SPA, ARE, RAP, SOP, POSE, PARE, PORE, SPORE, EAR, PEAR, APE, REAP, PEER, PEAR...........etc.

Pg. 53. (Your choice)

Pg. 55. Piano – Keyboard, Triangle – Percussion, Flute – Woodwind, Trumpet – Brass, Guitar - Strings

Pg. 57. SWIMS

Pg. 59.

Pg. 61. SEWING MACHINE

Pg. 69. $13.87

Pg. 71. A googol is written:
10,000,000,000,000,000,000,000,000,000,000,000,000,000,000,000,000,000,000,000,000,000,000,000,000,000,000,000,000,000,000,000,000,000,000

Pg. 73.

Pg. 75. 130 * 129 * 258 * 257 * 514 * 513

Pg. 77. $2.00 subtotal, $2.16 total

www.ingramcontent.com/pod-product-compliance
Lightning Source LLC
Chambersburg PA
CBHW081017040426
42444CB00014B/3250